Benretem Abdelouahab

A Theoretical Study of Centrifugal Pumps with Liquid-Solid Mixture

AF153206

Benretem Abdelouahab

A Theoretical Study of Centrifugal Pumps with Liquid-Solid Mixture

LAP LAMBERT Academic Publishing

Impressum / Imprint

Bibliografische Information der Deutschen Nationalbibliothek: Die Deutsche Nationalbibliothek verzeichnet diese Publikation in der Deutschen Nationalbibliografie; detaillierte bibliografische Daten sind im Internet über http://dnb.d-nb.de abrufbar.

Alle in diesem Buch genannten Marken und Produktnamen unterliegen warenzeichen-, marken- oder patentrechtlichem Schutz bzw. sind Warenzeichen oder eingetragene Warenzeichen der jeweiligen Inhaber. Die Wiedergabe von Marken, Produktnamen, Gebrauchsnamen, Handelsnamen, Warenbezeichnungen u.s.w. in diesem Werk berechtigt auch ohne besondere Kennzeichnung nicht zu der Annahme, dass solche Namen im Sinne der Warenzeichen- und Markenschutzgesetzgebung als frei zu betrachten wären und daher von jedermann benutzt werden dürften.

Bibliographic information published by the Deutsche Nationalbibliothek: The Deutsche Nationalbibliothek lists this publication in the Deutsche Nationalbibliografie; detailed bibliographic data are available in the Internet at http://dnb.d-nb.de.

Any brand names and product names mentioned in this book are subject to trademark, brand or patent protection and are trademarks or registered trademarks of their respective holders. The use of brand names, product names, common names, trade names, product descriptions etc. even without a particular marking in this work is in no way to be construed to mean that such names may be regarded as unrestricted in respect of trademark and brand protection legislation and could thus be used by anyone.

Coverbild / Cover image: www.ingimage.com

Verlag / Publisher:
LAP LAMBERT Academic Publishing
ist ein Imprint der / is a trademark of
OmniScriptum GmbH & Co. KG
Heinrich-Böcking-Str. 6-8, 66121 Saarbrücken, Deutschland / Germany
Email: info@lap-publishing.com

Herstellung: siehe letzte Seite /
Printed at: see last page
ISBN: 978-3-659-22263-4

Zugl. / Approved by: Annaba,University Badji-mokhtar Annaba, 2014

A THEORETICAL STUDY OF CENTRIFUGAL PUMPS
WITH LIQUID-SOLID MIXTURE

NOMONCLATURE

a	Acceleration	[m/s^2]
b	Wheel width	[m]
C	absolute speed	[m/s]
Cu	circumferential speed	[m/s]
Cr	radial velocity	[m/s]
C_V	Concentration by volume	[-]
C_p	Concentration peas	[-]
D	Wheel diameter	[m]
d_c	Diameter of the solid particles	[mm]
f	Friction	[-]
g	Acceleration of gravity	[m/s^2]
H	Delivery head of the pump	[m]
H_{th}	Head Euler	[m]
H_{dyn}	dynamic head	[m]
h	Load losses	[m]
Q	Pump flow	[m^3/h]
q	Volume solids	[m^3]
E	Linear momentum	[kg.m/s]
k	Coefficient of the quality of the blade	[-]
K_H	Reduction coefficient height	[-]
K_η	Reduction coefficient of performance	[-]
K_N	Correction coefficient of power	[-]
L	Length	[m]
M	Moment of force	[Nm]
m	Mass	[kg]
N	Power	[KW]
n	Speed	[tr/min]
P_m	Gauge	[kg/ cm^2]
P_v	vacuométrique pressure	[kg/ cm^2]
n_q	specific speed	$n_q = \dfrac{n.\sqrt{Q}}{H^{3/4}}$
ΔT	Work	[joules]
T	Temperature	[°K]
Re	Reynolds number	[-]
Γ	Traffic speed	[m/s]
z	Number of blade	[-]
Z	Height position	[m]
P	Pressure	[N/m^2]
W_c	Fall velocity of the particles	[m/s]
R_X	Trail	
R_y	Bearing	
U	Peripheral speed of the radius	[m/s]

Greek

ϖ	specific weight	[-]
ψ	Shrinkage coefficient	[-]
υ	kinematic viscosity	[m²/s]
P	Density	[kg/m³]
$\bar{\rho}$	relative density	[-]
δ	coefficient gauge	[-]
σ	Dimensionless coefficient of flow	[-]
ς	Dimensionless power coefficient	[-]
ω	angular velocity	[rad/s]
Φ	Viscous dissipation term	[W]
α	angle between $\vec{c}$ and $\vec{U}$	[°]
β	angle between $\vec{w}$ and - $\vec{U}$	[°]
η	Yield	[-]
τ	tangential stress	[N/m²]
μ	dynamic viscosity	[Pa.s]

INDICES

o.	Water
m.	for mixture
s.	for solid
1	wheel inlet
2.	Release of the wheel
v.	Volume
x,y,z	Cartesian coordinates
r,θ,z	cylindrical coordinates
i	gradient field

Abstract

Determining the individual characteristic of centrifugal pumps transporting solid liquid mixtures is of great importance in technological processes of pipeline transportation of various products such as iron ore, phosphate and coal. The selection of pumps for any transportation system depends mainly on previous experiments and/or or the execution of the pump with clear water. Several researchers have demonstrated that the operating system of centrifugal pumps depend on the behaviour of the mixture that is Newtonian or other.

The rheological properties are used to determine the selection of pumps to optimize the pressure drop i.e. the energy loss in the transportation; it is the transportation of solid liquid mixtures with a well defined concentration.

The purpose of this study is to show the influence of mechanical and physical properties of solids which are carried on the centrifugal pumps performance and provide a rational method to calculate the new mixtures characteristics based on those given for clear water and using the correction coefficients.

The first of the chapter is devoted to the rheological study of centrifugal pumps operating in mixture; two theories are proposed as the Euler implies that the absolute flow is permanent and axisymmetric at the wheel entrance, and possesses an infinite number of blades infinitely close. This arrangement, we cannot accept that admit the perfect fluid, ensures that the flow is permanent and axisymmetric at the wheel output. The vortex theory is closest to describe objectively the process of energy transfer for pumps operating in clear water. For mixtures, a theoretical study for evaluating additional losses caused by the presence of solid particles in the wheel rings.

The application of equations obtained in practice requires knowledge of the geometrical parameters of the wheel, the particle size of transported solid, the losses in the wheel and water mixture. All this leads to difficulties in applying these equations to determine the new characteristics of the pump operating in solid liquid mixture. In practice, empirical equations or graphs are used to recalculate the operating parameters of the pumps for specific conditions but the second part of the chapter is devoted to a literature on determining the characteristics of centrifugal pumps operating in solid liquid mixture where parameters that affect pump performance are shown: The mixture concentration, particle size and particle size of solid particles.

A THEORETICAL STUDY OF CENTRIFUGAL PUMPS WITH LIQUID-SOLID MIXTURE

1 Hydraulic characteristics of a mixture pump:

To determine the characteristics of excavation pumps, it should be noted that two theories are to be considered of fluid flow around the blades of the centrifugal pump impeller.

1.1 Application of the meshes theory of the operation process of pumps.

Meshes theory assumes essentially the absolute flow field of fluid is permanent and present at the entrance and exit of the wheel, a symmetry of revolution about the axis of the wheel [1]. Assuming that this requirement is met without much difficulty as regards the entry, it is reasonable to assume that it can be assuming that the wheel has an infinite number of identical blades, vanes infinitely thin and infinitely close from each other, symmetrically distributed about the axis of rotation.

Moreover, this theory assumes that the potential flow is without vorticity. The equation of the theoretical height of the pump characterizes the process of transmission of the power applied to the pump shaft to the flow of liquid without loss of energy. Whereas in the parallelograms speeds of the fluid particles at the entrance and exit of the impeller Figure.1, using the moments theorem of change amount of motion relative to the axis of the machine, the moments equation M can therefore be obtained as follows:

$$M \ = \ \frac{d\,E}{d\,t} \qquad (1)$$

En by applying the continuity equation for input sections and output channels of the impeller the equation of the theoretical flow created by the pump is:

$$Q_v = D_2 . b_2 . \pi . C_{r2} . \psi_2 \qquad (2)$$

Where ψ_2 is a factor that takes into account the reduction of the area output by the blades. One can determine the equation of the theoretical height created by the pump operating in clear water [2]:

$$H_{th\infty} = \frac{1}{g}\left(U_2^2 - \frac{U_2 . ctg\, \beta_2 . \, Q_{th}}{D_2 . \pi . b_2 .} \right) \qquad (3)$$

By analyzing the equation (3), it can be seen that ψ_2 coefficient is unknown. Moreover the meshes theory does not give an answer about the influence of the number of blades on the pump characteristic. If pressure drops of a real machine are determined by experience, the theoretical height created by the latter with an infinite number of blades is obtained. Note that it is less than that calculated for an infinite number of blades. This difference can be explained by the discrepancy between meshes theory and of the actual process in the impeller.

According to this theory, we take as a basis the flow symmetry in the channels of the wheel with constant speeds in any point on the circumference that is to say that the resultant force on the shaft of the pump is zero and the wheel does not create pressure but in reality with a finite blades number, the pressure upstream of the blade is greater than that which causes forward energy conversion in the impeller.

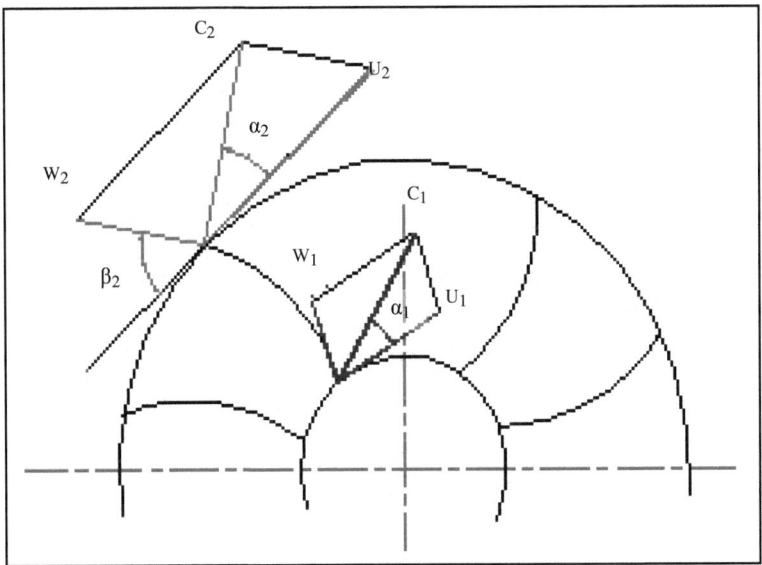

Figure.1 *Parallelogram speeds at the entrance and exit of the wheel*

The impeller transmits to fluid a rotational movement which gives different pressures and different velocities along the blade that is why the relative flow in the channels of the wheel does not have the fillets shape but is done with vortices. The swirling motion to the input channels has a negative direction with a low speed and the relative velocity of the vortex at the entrance of the channels has an opposite direction compared to that of the driving speed so we will have a deviation of flow of the wheel and the circumferential speed decreases. The theory of vortices allows in presenting a clear physical meaning of the mutual forces action between the blades and the liquid. These forces determine the character of the flow movement and structure. Consider the basis of this theory and apply it on the movement of bulk flow in the impeller.

1.2 Application of the of vortices theory on the flow process of the liquid in the impeller.

a) Circulation of the outline: The circulation of the velocity vector is the kinematics characteristic of turbulent motion of fluid particles around any fixed instantaneous axis or moved. It represents the work of the velocity vector from the closed contour, i.e. the movement is analogous to the mechanical work but here is the strength is shown by speed

$$\Gamma = \oint v.\cos\varphi.ds \qquad (4)$$

Where: v- Flow velocity in the given point of the contour.

ds - Elementary contour length.

φ - Angle between the velocity vector v and the element ds.

The simplest case is the outline round figure.2

$$\Gamma = 2\pi.R.v = 2.\pi.R^2.\omega = 2\int_S \omega ds$$

With ds the section.

The circulation of the velocity vector depends only on the value of the vortices that cause it and it is not determined by the shape of the closed contour. The movement of the contour is equal to the sum of the internal circulation of the contour.

$$\Gamma = \sum_{i=1}^{n} \Gamma_i.$$

b) Traffic flow around a body:

Consider the flow of a plate by a flow during normal movement, figure 3. Speed meshes decreases as it approaches the plate that is why the pressure increase follows Bernoulli's law [3, 4]. At the periphery the flow shrinks and subsequently the speed increases and the pressure decreases. After crossing the plate was the flux relaxation and meshes are a path of alternating vortices due to low pressure. The vortices that cause pressure reduction are due to the change in mass flow of liquid. The plate has a bad flow shape, there is a large difference between the pressures upstream and downstream, due to that the fluid resistance force R which acts contrary to the movement. In order to obtain this force, the momentum theorem can be applied. A portion of fluid particles begins to move with the plate, and the mass of these particles for a unit of time is:

$$m = \rho.S.v_1 \qquad\qquad (5)$$

Where: ρ - fluid density, S - surface of the plate, v_1 -particle velocity. The speed is proportional to the speed v of the plate

$$v_1 = c_1.v$$

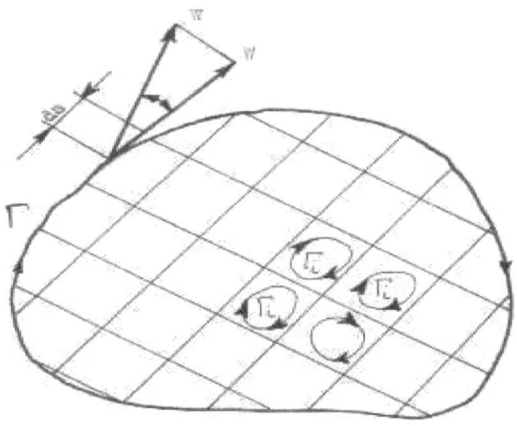

Figure.2 *Flow diagram according to an elementary surface*

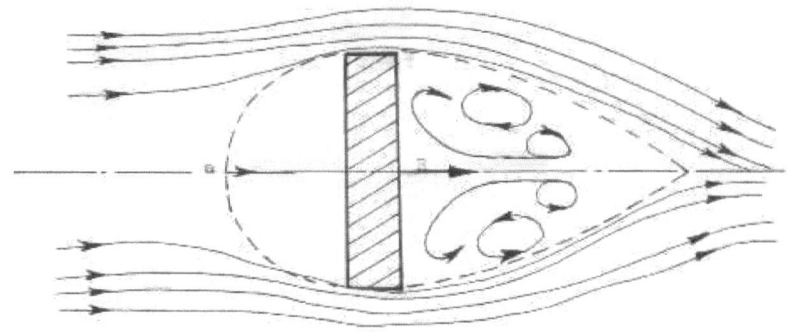

Figure.3 *A plate movement by a flow*

And therefore the mass momentum of the fluid during the time unit must be equal to the resistance R:

$$R = mv_1 = mC_1.v \tag{6}$$

According to this equation, it can be deducted that:
- The value of R is proportional to the descent of the fluid,
- It depends on the shape of the body C_1, and is proportional to its cross-section S.
- It is directly proportional to the square of the velocity v.
We can write:

$$R = \frac{1}{2}\rho.\partial.S.v^2 \tag{7}$$

Here we take:

$$\partial = 2.C_1$$

This formula is essential to establish the action of forces on the turbine blades of a pump on the liquid and thus on the solid particle during hydraulic transport.

c) The flow profile by the asymmetric flow:
To approach to the turbine, consider the flow profile of the fin by the asymmetric flow Figure.4, speed W_C depends on the shape of the profile of the fin and the attack angle α.
Next Bernoulli's pressure increases in the work side of the blade. The pressure difference is an essential requirement for the transmission of mechanical energy to the liquid flow. The operation of the engine without moving traffic is impossible. The circulation of the velocity vector determines the quantitative value of vortex motion.

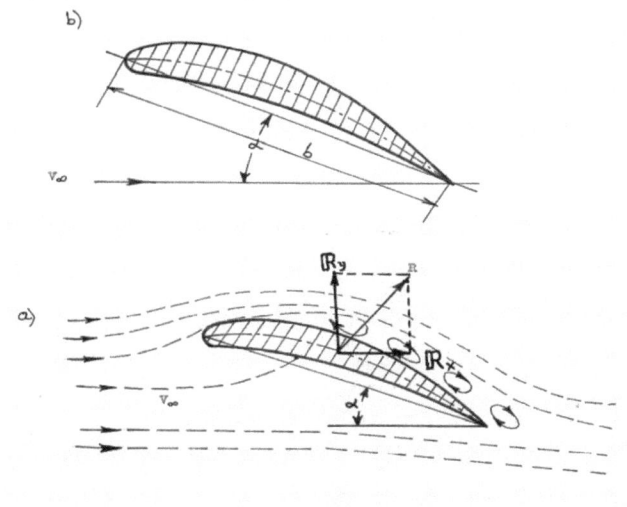

Figure.4 *Movement of a body by a flow of liquid*

13

The blade strength on the flow of liquid is determined by the formula of Kutta-Joukovsky [ ] 5.6 which gives the relationship between force and increase the flow velocity vector around the contour of length I and width b; figure.4.b.

$$R_y = \rho.v_\infty.\Gamma.L \qquad (8)$$

The force direction R$_y$ is determined by rotating the velocity vector V∞ unlike a right angle to the direction of movement of traffic. Using equations (7) and (8) we can write:

$$\rho.v_\infty.\Gamma.L = \partial.\rho.S\frac{v_\infty^2}{2}$$

This equation gives the relationship between the calculated value Г C and the value obtained experimentally for different profiles. The force of frontal resistance is obtained experimentally. The relationship between the lifting force (lift) and R$_y$ resistance force R gives us the coefficient of the quality of blade or fineness.

$$k = \frac{R_x}{R_y} = tg\ \alpha \qquad (9)$$

1.3. Analysis of the impeller operation by applying the vortices theory.

The relative motion of the liquid in the channels of the impeller[5] by three main movements figure 5, is presented:
- Movement in the fixed grid with the relative speed W$_f$.
- Swirling motion within the channels caused by the rotation of the grid with the relative speed W$_t$.
- Circular motion around the blades with the relative velocity W$_p$.
In this case the relative speed of the liquid in the channels is equal to the sum

of the three speeds:

$$W = W_f + W_t + W_c$$

The relative speed W_f decreases at the output due to the increase in the channel section. The speed W_t appears if we assume that the output and input channel are closed and filled completely with liquid. The tourbillion in the wheel takes place because of the inertia of the liquid rotating with the angular velocity ω.

The direction of the swirling motion is the same as for the driving speed U to the input of the wheel. The speed value of the vortex increases with distance from the rotation centre. The existence of the swirling motion creates the speed difference on the surface of the blades.

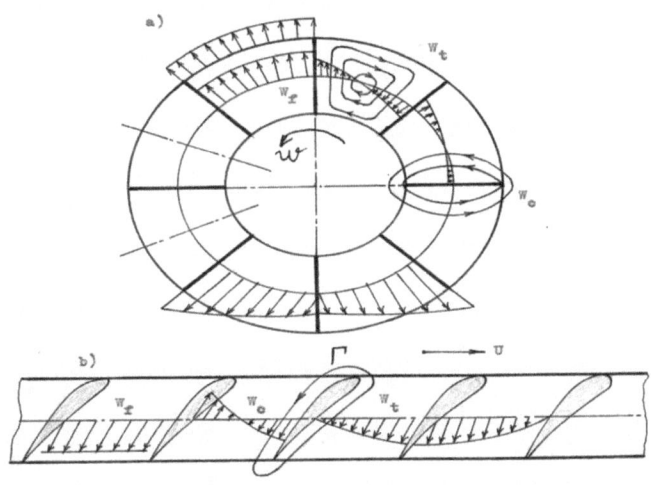

Figure.5 *Relative movements in an impeller of a centrifugal pump*

Consider the line abcd as a flow outline in the centrifugal pump figure.6. The lines a b and c d contour have opposite relative velocities. In this case the movement of such lines is as follows:

$$\Gamma ab = -\Gamma cd$$

The circulation of the velocity vector around a blade is:

$$\Gamma a = \Gamma ad + \Gamma bc + \Gamma cd + \Gamma ad$$

Moreover we know that for a centrifugal pump:

$$\Gamma a = t_2 Cu_2 - t_1 Cu_1 \tag{10}$$

t_2 of the grid.

Where $\qquad t_1 = \dfrac{2.\pi.R_1}{z} \quad ; \quad t_2 = \dfrac{2.\pi.R_2}{z}$

z - Number of blades in the impeller

So for all centrifugal pumps we obtain:

$$\Gamma a_{cent} = \frac{2\pi}{z}(R_2 Cu_2 - R_1 Cu_1)$$

And for a number of blades z we have:

$$\Gamma a_{cent} = \Gamma ac.z = 2\pi(R_2 Cu_2 - R_1 Cu_1) \tag{11}$$

The analysis of the equation (11) shows that the movement around the grid does not depend on the nature of the fluid.

The theory that has just announced, the circulation around the blade causes the mutual forces between the blades and the liquid flow. The work of these forces is the energy transmitted by the wheel directly to the liquid. By

16

analyzing the movement of fluid in the grid we notice that the circulation around each blade consists of two circulations: caused by the vortex inside the channel and caused by the flow of blade with the special profile.

$$\Gamma ac = \Gamma 1 \pm \Gamma 2$$

Here, the sign (+) to the blades bent forward and the minus sign to those bent backwards.

For the fundamental equation of the turbo machine, equation (11) is multiplied by the angular velocity.

.

$$\Gamma w = 2\pi (U_2 Cu_2 - U_1 Cu_1) \tag{12}$$

The part in brackets is the theoretical height determined by Euler [7].

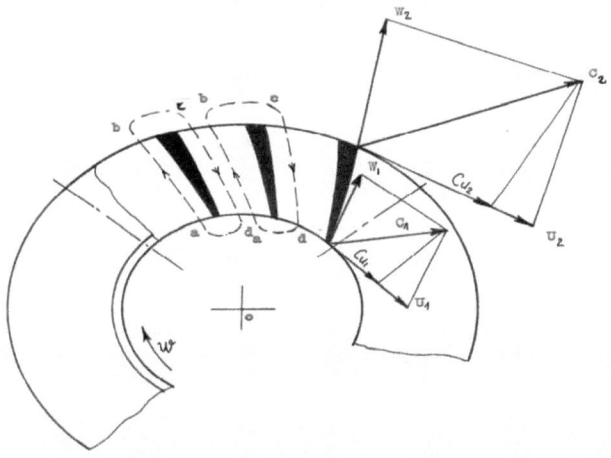

Figure.6 *Circular flow around a blade of a radial centrifugal impeller*

$$H_{th} = \frac{\Gamma . \omega}{2\pi . g} \qquad (13)$$

It can be concluded that the theoretical height does not depend on the liquid nature, but it is a function of circulation number of blades and the rotation speed.

It can also be said that the theory of vortices does not use new principles we only change this expression $R_2 Cu_2 - R_1 Cu_1$ by $\frac{\Gamma a}{2\pi}$ which gives us the connection with aerodynamics that lightens the natural phenomena that occur in the internal channels of the impeller of a centrifugal pump.

The specific operating condition of centrifugal pumps excavation requires a special design of the impeller. Specifically, the number of blades varies from 1 to 6, this causes inequality between sections of the input and output channels and hence the separation of the flow figure7. The application of the vortices theory, the correction coefficients of the theoretical height created by the pump, can be determined.

1.4. Determination of correction coefficients of theoretical height of an excavated pump operating in water.

The presence of separation zones changes the character of the blades flow by the liquid compared to the normal pump [8]; this is why the application of the coefficients of the circumferential speed of the shrinkage and to determine the theoretical head of the pump is impossible. Moreover it is impossible to use the theory of rotating grids for these wheels if the zones of separation are unknown. The theoretical characteristic of the pump with a finite number of blades is determined by the equation [7]:

$$H_{th_z} = \frac{1}{g}\left[\left(U_2^2 \mu - \frac{Q_{th} U_2}{\pi . D_2 b_2 . \psi_2} ctg\right) - Cu_1 U_1\right] \qquad (14)$$

18

Where: D_2 - outside diameter of the impeller, m.

            μ - Correction coefficient of the number of blades.

            b_2 - width of the wheel; m.

            ψ_2 - a factor that takes into account the reduction of the exit
area by the blade.

At the present time, the approximate methods are used to determine the Stodola separation zones [3] assumed that the relative flow between the wheel blades is composed of two flows: normal flow and with vortex figure.7.b. The flow with eddies has a negative direction with respect to the rotational speed of the impeller. Consider the relative movement of the liquid in the wheel taking into account the rotation of any system with the angular velocity ω and analysing μ and ψ_2 coefficients. In this case, the relative flux as said is divided into two: one is the flow in the case of finite blades number, the other created by the vortices. The boundaries of the vortex surfaces are arcs of the blades and the circumferences of input and output of the impeller. The existence of the vortices causes the birth of increased speeds to the input and output and ΔW_1 ΔW_2. In this case the parallelogram speeds

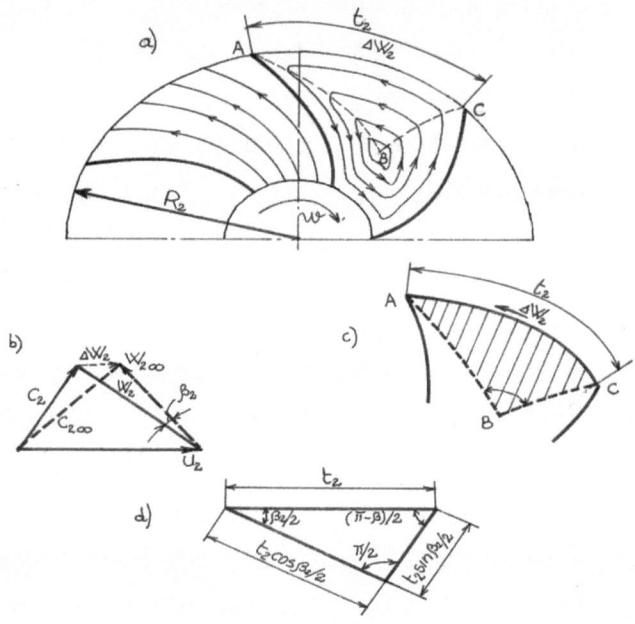

Figure.7 *The flow diagram of the flow through the blades of the pump impeller*

change compared to that for $z = \infty$ and the value of β_2 and decreases. By applying the calculation scheme for the flow with follow on area between the blades, vortices will be limited by the surface of the blade and the separation zone at the rear of the adjacent blade Fig.7 a, b. To simplify the determination of the coefficient γ, it is assumed that the line of the separation zone is equidistant to the face of the blade which is actually ab cb. If the distance between the axes of adjacent blades is equal to:

$$t_2 = \frac{2.\pi.R_2}{z}$$

20

The length of the vortex line along the circumference at the output is equal to $t_2 \psi_2$. Noting that the flow velocity vector along the contour is determined by Stodola, this is why we the triangle ABC where BC and AB lines are normal lines of vortices while is considered.

$$AC = t_2 \psi_2.$$

According to Stokes theorem, the flow velocity vector along the contour ABC is:

$$\Gamma = 2.\omega.S_{abc} = \frac{2.\pi.R_2}{Z} \Delta W_m.\psi_2 \qquad (15)$$

Where: ΔW_m - average speed of the vortex along the line AC.

Traffic speed along the lines AB and BC is zero because of the normal direction to the vortex lines.

By changing the surface of the triangle ABC by that of a right triangle we have:

$$2.S_{abc} = t_2.\psi_2.\cos(\frac{\beta_2}{2}) \sin\left(\frac{\beta_2}{2}\right).t_2\psi_2$$
$$= t_2^2.\sin \beta_2.\left(\frac{\psi_2^2}{2}\right) \qquad (16)$$

Equations (15) and (16) we get:

$$w_m = \omega.\frac{t_2\psi_2}{2} \sin \beta_2 = \omega.\psi_2\frac{\pi.R_2}{z} \sin \beta_2 =$$
$$= \frac{U_2.\pi.\sin \beta_2}{z} \psi_2$$

In figure 8, we can write:

$$Cu_{2\infty} = U_2 - Cr_2.ctg \beta_2 \qquad (17)$$

21

Taking into account the presence of vortices and on the narrowing of the flow in the blades, the speed component of the output will be:

$$Cu_2 = Cu_{2\infty} - \Delta W = U_2 - \frac{Cr_2.ctg\beta_2}{\psi_2} - \frac{U_2.\pi.\sin\beta_2.\psi_2}{z}$$

$$Cu_2 = U_2(1 - \frac{\pi.\psi_2.\sin\beta_2}{z}) - \frac{Cu_2.ctg\beta_2}{\psi_2}$$

(18)

Using equations 17 and 18, we deduce:

$$\mu = 1 - \frac{\pi.\sin\beta_2}{z}\psi_2$$

(19)

Using equation (19), it can be seen that it affects the coefficient and the theoretical speed of the pump. Now, determine the influence of the wheel parameters on the coefficient ψ_2.

From equation18, we can write:

$$\frac{Cr_2.ctg\beta_2}{\psi_2} + \frac{U_2.\pi.\psi_2}{z}\sin\beta_2 = U_2 - Cu_2$$

It can be seen that the left side of this equation has a minimum for a well defined value of shrinkage coefficient ψ_2. The condition of the flow formation separation for a maximum speed Cu_2 is presented by the following differential equation:

$$\frac{\partial}{\partial\psi_2} = (\frac{Cr_2.ctg\beta_2}{\psi_2} + \frac{U_2.\pi.\sin\beta_2}{z}\psi_2) = 0$$

$$\psi_2 = \sqrt{\frac{Cr_2.z.ctg\beta_2}{\sin\beta_2.U_2.\pi}}$$

(20)

To specify the value of the coefficient y, we determine the angle correction β_2 figure.8.

The curvilinear grid of blades is shown by the lines CF and and the flow separation occurs from the cb line. In this case the normal to the flow line moves to the mass of the vortex (I) and the centre of rotation is 0 ′. Suppose that the angle of b0′d is equal to 90 ° in this case the angle 0db0′db is less than 0db which means that one must have $\beta_2 - \Delta\beta_2$ instead of β_2 in formula (19). In considering the triangle the cba and curved b a c, we can write:

$$\frac{180^0 - \beta_2}{2} + \frac{\theta}{2} - \frac{180^0 - \beta_2 + \theta/2}{2} = \frac{\Delta\beta_2}{2}$$

$$\frac{tg(90^0 - \beta_2 + \theta)}{tg(90^0 - \beta_2)} = \frac{c.h}{a.h}$$

By putting ah = α and ac=ψ α$_2$ t$_2$ we have:

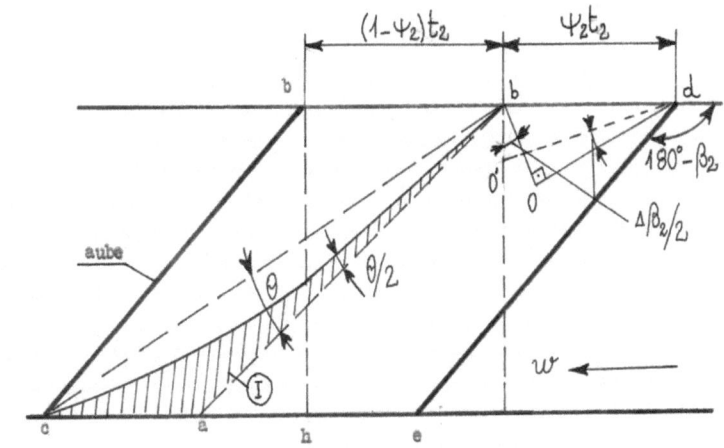

Figure.8 *Calculation scheme of the β$_2$ angle*

$$\frac{tg(90^0 - \beta_2 + \theta)}{tg(90^0 - \beta_2)} = \frac{(1-\psi_2)t_2 + \alpha\psi_2}{\alpha\psi_2} = \frac{1}{a}$$

where:

$$a = \alpha\psi_2 \frac{1}{(1-\psi_2)t_2 + \alpha\psi_2}$$

and

$$tg\ \theta = \frac{(1-a)}{tg\beta_2(a + 1/tg^2\beta_2)}$$

According [6] for the operating wheels in solid-liquid mixture α is between 0.7 and 1.5 and the variation of the angle α affects $\Delta\beta_2$ (about 2 °). Taking $\alpha = 1$ and $\psi_2 = a$ we have:

$$tg\theta = \frac{1-\psi_2}{tg\beta_2(\psi_2 + \frac{1}{tg\beta_2})} = \frac{tg\beta_2 - \psi_2 tg\beta_2}{1 - \psi_2 tg^2\beta_2}$$

and as $\Delta\beta_2 = \frac{\theta}{2}$:

$$\mu = 1 - \frac{\pi\sin(\beta_2 - \Delta\beta_2)}{z}\psi_2 \qquad (21)$$

The theoretical height of the excavated pump operating with water $Cu_1 = 0$ is:

$$H_{th} = \frac{1}{g}\left[U_2^2(1 - \frac{\pi\sin(\beta_2 - \Delta\beta_2)\psi_2}{z}) - \frac{Q_{th}.ctg\beta_2}{2\pi.b_2.\psi_2.R_2}\right] \qquad (22)$$

This equation is valid for machines with velocity from 45 to 200 the number of blades 2 is 16 from 1 to 6 and the angle of inclination β_2 is from 16 to 35 °.

2 Parameters influencing the characteristics of centrifugal pump

2.1. Theoretical height of the pump operating as a mixture:

Let the velocities of the solid particles and liquid at the outlet of the wheel are different and denote the absolute velocities of the circumferential components to the speed of the impeller by solid particles and that of the liquid $Cu_2.s$ by $Cu_2.o$. The change in the movement amount of the mass of the solid and liquid unit of time is equal to:

$$M = Q_s.\rho_s.Cu_{2,s}.R_2 + Q_o.\rho_o.Cu_{2,o}.R_2 \qquad (23)$$

Where　　　Q_s and Q_0 flow of solid and liquid

ρ_s ρ_o density of the solid and the carrier liquid

The change in time of the momentum of the mass movement of the mixture is equal to the forces applied to the blade of the impeller. The pressure forces at the entrance and exit of the wheel have a direction perpendicular and therefore they do not cause the moments about the axis of the impeller. Multiplying the equation by the angular velocity we get:

$$M.\omega = P = Q.z.\rho_s.U_2.Cu_{2,s}$$
$$Q_o.\rho_o.U_2.Cu_{2,o} = H_{th.m}.Q_m.\rho_m.g$$

Q_m et ρ_m are respectively the density of the charged mixture. And the theoretical height created by the pump operating with mixture will be:

$$H_{th.m} = \frac{U_2.Cu_{2,s}.Q_s.\rho_s}{g.Q_m.\rho_m} + \frac{U_2.Cu_{2,0}.Q_o.\rho_o}{g.Q_m.\rho_m} \qquad (24)$$

25

By appointing C_v mass concentration, the theoretical height becomes:

$$H_{th.m} = \frac{1}{g} C_v.U_2.Cu_{2.s} + (1-C_v)U_2.Cu_{2.o} \qquad (25)$$

With:

$$C_v = \frac{m_s}{m_m}, 1-C_v = \frac{m_o}{m_m}.$$

Here m_o, m_s, m_m are respectively the mass of water, and solid mixture.
The height created by the pump for water and for the mixture is different and are equal only if the circumferential speed of solid $Cu_{2.s}$ is equal to that of water $Cu_{2.o}$.

$$H_{st} = H_{th} - H_{depr} - h_{pr} \qquad (26)$$

Where: h_{pr} – pressure losses in the impeller.
In equation. (24) and replacing the height by its expression is the static head as follows:

$$H_{st.m} = C_v \frac{U_2.Cu_{2.s}}{g} + (1-C_v)\frac{U_2.Cu_{2.o}}{g} - C_v \frac{C_{2.s}^2 - C_1^2}{2g} - (1-C_v)\frac{C_{2.o}^2 - C_1^2}{2g} - h_{pr} \qquad (27)$$

Consider the parallelogram of velocities at the exit wave and the wheel for the liquid and the solid figure 9, we can write:

$$C_{2.s}^2 + W_{2.s}^2 - (U_2 - Cu_{2.s})^2 = W_{2.s}^2 - U_2^2 + 2U_2.Cu_{2.s} \qquad (28)$$

$$C_{2.s}^2 = W_{2.s}^2 - U_2^2 + 2U_2 Cu_{2.s} \qquad (29)$$

$$C_1^2 = W_1^2 - U_1^2 \qquad (30)$$

Where: $C_{2.s}$ and $W_{2.s}$ and are respectively the absolute and relative velocities of the solid. And therefore:

$$H_{st.m} = \frac{C_v.U_2.Cu_{2.s}}{g} + (1-C_v)\frac{U_2.Cu_{2.o}}{g} - \frac{C_v}{2g}\left(W_{2.s}^2 - U_2^2 + 2.U_2.Cu_{2.s} - W_1^2 + U_1^2\right) - $$
$$\frac{1-C_v}{2g}\left(W_{2.o}^2 - U_2^2 + 2.U_2.Cu_{2.o} - W_1^2 + U_1^2\right) - h_{pr}$$

Où :

$$H_{st.m} = \frac{U_2^2 - U_1^2}{2g} - \frac{W_1^2 - W_{2.o}^2}{2g} - C_v.\frac{W_{2.s}^2 - W_{2.o}^2}{2g} - h_{pr} \qquad (31)$$

It can be seen that the static head created by the impeller operating in mixture changes compared to running water, this change is due primarily to the increase in dynamic height and loss of additional charges.

2.2 The additional pressure drop in the impeller.

For the pressure drop, consider the movement of solid particles relative to the inertial coordinate system figure (10) which rotates together with the impeller. Introduce the Coriolis force and that centrifugal to examine the relative movement of solid particles and the relative velocity of the liquid W_0, so the normal component of acceleration in the relative movement [8] is equal to:

$$a = 2.W_o.\omega - \omega^2.R.\cos\varphi - \frac{W_o^2}{R_f} \qquad (32)$$

R - Distance between the centre of the wheel and the solid particle, m
φ - angle between the normal to the line of flow and the radius vector of the given point.
R_f - Curvature Flow. It is known that the solid particles settle under the action of the force of gravity in pipes during transport. In the channels of the pump,

speed of the solid is almost the same for water but the acceleration is greater, this causes the separation of solid particles move relative to the liquid which causes the appearance of normal resistance and therefore the increase in additional losses. Consider the relative movement of the solid particles relative to the liquid. The Coriolis force is normal to the trajectory of particle motion that is why it does not depend on work. The projection of the centrifugal force F is tangent to the path of movement.

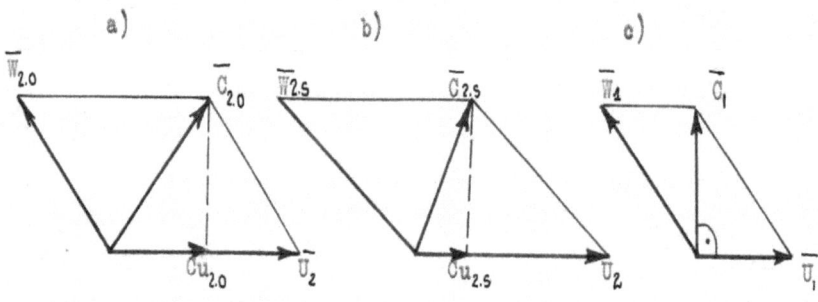

Figure.9 *Parallelogram speeds of water a) for the solid and b) at the exit and entry of the wheel*

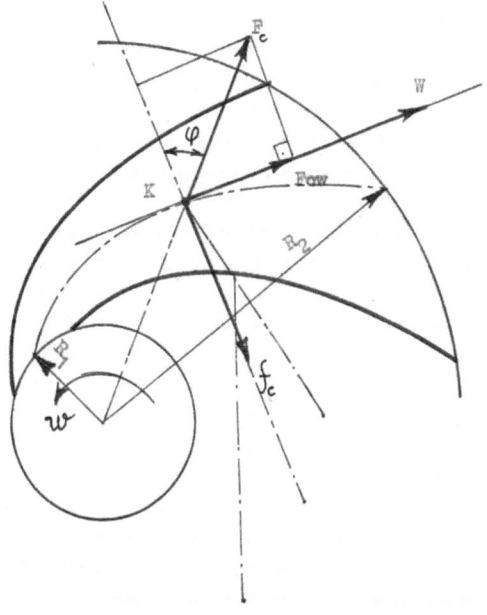

Figure.10 *Diagram of forces acting on a solid particle*

$$q.\rho_o.\omega^2.R.\sin\varphi = q.\rho_s.\omega^2.R.\frac{dR}{dS} \qquad (33)$$

Where: q - volume of solid particles,

ρ_s - solid density, kg/m^3

dS - path length, m

The work done by the centrifugal force during the movement of the particles along the path is:

$$\int_{R1}^{R_2} q.\rho_s.\omega^2 R \frac{dR}{dS}.dS = q.\rho_s \frac{\omega^2.R_2^2 - \omega^2.U_1^2}{2} = q.\rho_s \frac{U_2^2 - U_1^2}{2}$$

R_1 and R_2 are respectively the radii of input and output of the impeller, m.

The work expended by the impeller to move the particles is: $q(P_2 - P_1)$.

Denote the work of normal resistive forces of a single particle at the entrance and exit of the wheel:

$$\rho_s.q \frac{W_{2.s}^2 - W_1^2}{2} + q(P_2 - P_1) = q.\rho_s \frac{U_2^2 - U_1^2}{2} - \Delta T$$

Using equation (31) we get the work forces of resistance spent during normal particle motion:

$$\Delta T = q(\rho_s - \rho_m)\frac{U_2^2 - U_1^2}{2} - q.\rho_m \frac{W_1^2 - W_{2.o}^2}{2} + q.S.\rho_m \frac{W_{2.s}^2 - W_{2.o}^2}{2} + q.\rho_m.g.h_{pr} - q.\rho_s \frac{W_{2.s}^2 - W_1^2}{2}$$

$$\Delta T = q(\rho_s - \rho_m)\left[\frac{U_2^2 - U_1^2}{2} + \frac{W_1^2 - W_{2.o}^2}{2}\right] - q(\rho_s - \rho_m.C_v) + \frac{W_{2.s}^2 - W_{2.o}^2}{2} + q.\rho_m.g.h_{pr} \quad (34)$$

The losses in the impeller may arise as the sum of water losses $h_{r.o}$ and additional losses Δh_r due to normal resistance of solid particles $h_{r.o}$ moved relatively to the liquid flow:

$$h_{pr} = h_{r.o} + \Delta h_r \quad (35)$$

Small particles following the intense agitation remaining in suspension despite the existence of the great acceleration. If we denote the volume concentration of the mixture by C_v, the small particles displaced by the action of turbulent motion by C_{v1} and C_{v2} that of large particles in the mixture, we have:

$$C_v = C_{v1} + C_{v2}$$

The amount of particles displaced relatively to the flow of liquid during the unit time is equal to:

$$\frac{C_{v2} \cdot Q_m}{q}$$

The work expended to overcome the natural resistance of particles per unit mass of the mixture represents additional losses:

$$\Delta h_r = \frac{C_{v2} \cdot Q_m}{q} \frac{\Delta T}{Q_m \cdot \rho_m \cdot g} = \frac{C_{v2} \cdot \Delta T}{q \cdot \rho_m \cdot g} = C_v (1 - \frac{C_{v1}}{C_v}) \frac{\Delta T}{q \cdot \rho_m \cdot g} \qquad (36)$$

Equations (35) and (36), we obtain the value of additional losses:

$$\Delta h_r = \frac{C_v}{\rho_m \cdot g} (1 - \frac{C_{v1}}{C_v}) \left[\begin{array}{l} (\rho_s - \rho_m)(\frac{U_2^2 - U_1^2}{2} + \frac{W_1^2 - W_{2.o}^2}{2}) - \\ (\rho_s - C_v \cdot \rho_m)(\frac{W_{2.s}^2 - W_{2.0}^2}{2}) + \rho_m \cdot g(h_{r.o} + \Delta h_r) \end{array} \right]$$

So:

$$\Delta h_r = \frac{C_v (1 - \frac{C_{v1}}{C_v})}{1 - C_v (1 - \frac{C_{v1}}{C_v})} \left[\begin{array}{l} \frac{\rho_s - \rho_m}{\rho_m} (\frac{U_2^2 - U_1^2}{2g} + \frac{W_1^2 - W_{2.o}^2}{2g}) - \\ - \left(\frac{\rho_s - C_v \cdot \rho_m}{\rho_m} \right) \frac{W_{2.s}^2 - W_{2.0}^2}{2g} + h_{r.o} \end{array} \right] \qquad (37)$$

Since the values of terms $\frac{W_{2.s}^2 - W_{2.o}^2}{2g}$ and $\frac{W_{2.s}^2 - W_{2.o}^2}{2g}$ are negligible and we can simplify the formula (37) as follows:

$$\Delta h_r = \frac{C_v (1 - \frac{C_{v1}}{C_v})}{1 - C_v (1 - \frac{C_{v1}}{C_v})} \left[\frac{\rho_s - \rho_m}{\rho_m} \frac{U_2^2 - U_1^2}{2g} + h_{r.o} \right] \qquad (38)$$

Note that the additional loss of pressure in the pump operating as a mixture depending on the dimensions of solid particles and the amount of small particles in the mixture. The calculation of losses is made with the separation of the large and small particles. The critical diameter is called a size limit of particles with all particles smaller than the latter are moved with the liquid and the others have a relative motion to the liquid. To establish the equation of the critical diameter of particles, consider the movement of the mixture contained in a channel (11).

1-1 be on the 1-1′ surface in unit volume in the amount of N_z particles and that on the 2-2′ surface located by the distance 1 of 1-1′. The particles velocity in the channel is equal to W′. The root mean square of the velocity pulsations in the channel is equal to V′. Suppose that the mixture process is done by speed pulses V′

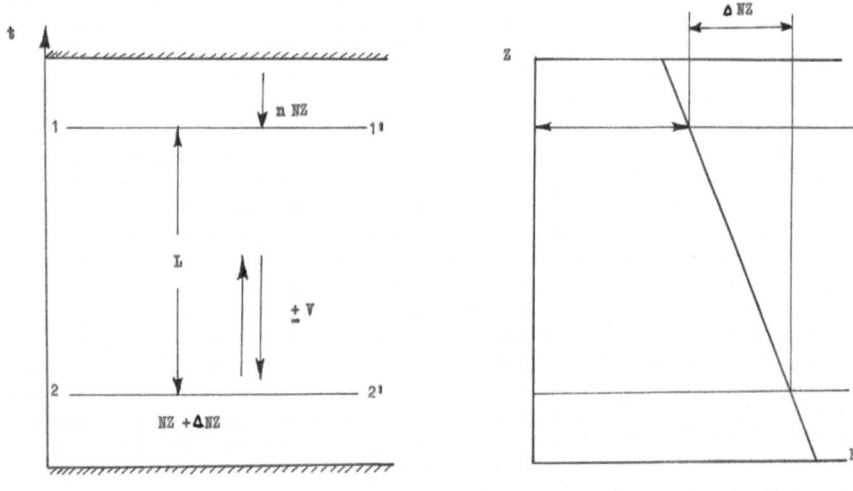

Figure.11 *Movement of the mixture in a channel*

and their settling due to the falling speed W'. If this process is taken as permanent we can write:

$$(V'+W')N_z = V'(N_z + \Delta N_z)$$

This means that during the unit time in the layer of 1-1' the amount of outgoing particles is equal to those entering particles in the layer 2-2'.
Using equation (23) and the law of solid particles distribution in the surface along the field direction of acceleration we can write the following relation:

$$\frac{\Delta.N_z}{N_z} = \frac{W'}{V'}$$

The settling particles velocity at rest W' depends upon driving forces and field acceleration forces. This ratio ΔN_z / N_z has to be done so that the particles distribution law of same direction is equal to that given in the section, we take in this constant case and the criterion that characterizes the distribution law is the ratio of W'/V ' i.e. the similarity criterion of solid particles distribution. Under the action of acceleration in all channels, only the small particles which remain.
The settling velocity W' of particles with a diameter d_c displaced under action of the acceleration is determined by the following relationship:

$$\frac{\pi.d_c^3}{6}(\rho_s - \rho_o)a = \lambda.\rho_o \frac{\pi.d_c^2}{2}.\frac{W'^2}{2} \qquad (39)$$

The resistance coefficient λ can be taken approximately according to the Stokes law.

$$\lambda = \frac{24}{\text{Re}} = \frac{24.v}{W'.d_c}$$

Here v: - kinematics viscosity of the liquid.

And the rate of fall:

$$W' = \frac{\rho_s - \rho_o}{\rho_o} \frac{d_c^2}{18.v} a \quad ; \text{m/s}$$

And the similarity criterion of distribution is:

$$\frac{W_c'}{V'} = \frac{\rho_s - \rho_o}{\rho_o} \frac{d_c^2.a}{18.v.V'} = const \qquad (40)$$

Here: W'_c - critical velocity of particles

Suppose that V' is proportional to the average relative velocity W_m in the blades i.e. $V' = \alpha W_m$. And as we know for similar pumps, the speeds are proportional to ωD and acceleration to $\omega^2 D$ and therefore the ratio a/W_m is proportional to ω.

$$\frac{\rho_s - \rho_o}{\rho_o} \frac{d_c^2.a}{18.v.V'} = \frac{\rho_s - \rho_o}{\rho_o} \frac{d_c^2.n}{18.v.\alpha}$$

And equation (3.2.18) becomes:

$$\frac{W_c'}{V'} = \frac{d_c^2.\omega}{18.v.\alpha} \frac{\rho_s - \rho_o}{\rho_o} = const$$

Where:

$$\frac{d_c^2 \sqrt{\omega}}{\sqrt{\upsilon}.\alpha} = \sqrt{\frac{W_c'}{V'} . \frac{18.\rho_o}{\rho_s - \rho_o}}$$

Assuming that the solid particles of same size will settle at the same time ($W'_c/V'= 1$) and α=15% the value of the critical diameter will be:

$$d_c = 1,65 \sqrt{\frac{\upsilon}{\omega}} . \sqrt{\frac{\rho_s . \rho_o}{\rho_s - \rho_o}} \qquad ; mm \qquad (41)$$

The order of determining the loss of additional charges is as follows:
* The critical diameter of solid particles from equation (41).
* The concentration of small particles which have a diameter less than that using the critical particle size distribution of the solid.
* The value of losses in the wheel according to the equation (37) below:

$$h_{r.o} = H\left[(1-\eta_r)/\eta_h\right]$$

Where: η_r - the action wheel efficiency depends on the number of blades.

η_h - hydraulic efficiency which is determined by the following equation:

$$\eta_h = \frac{n_s}{n_s + 24} . \eta_r$$

Where: n_S – the pump specific speed.

The height created by the excavation pump operating as a mixture of S concentration is equal to:

$$H_{pm} = (H_o - \Delta h_r) \rho_m$$

Here: H_0 - Height of excavation pump water.

Δh_r - are the additional losses determined according to equation (38).

ρ_m - Density of the mixture. Kg/m^3

The theoretical equations of excavation pump considered in this chapter allow in knowing the influence of solid particles, the number of blades and the construction of the impeller on the pump operating characteristics of a mixture. But it should be noted that the establishment of these equations is performed with assumptions.

The application of these equations requires knowledge of the geometric parameters of the impeller, the solid particle size, pressure losses in the impeller in water and in mixture ... etc. All this leads to difficulties in applying these equations to recalculate the pump characteristics. In practice, empirical equations are used in recalculation of graphs but for concrete conditions. Let us analyze these equations to find the method closest to our conditions.

3 Analysis of correcting characteristics methods of centrifugal pumps

Because of their great advantages: flow consistent, easy maintenance, stability, centrifugal pumps are widely used in hydraulic transport systems of solid particles. The selection of pumps in transport depends mainly on the results of experiments already conducted on the characteristics of clear water pumps [7.21, 22, 23]. When solid particles are introduced into the liquid, the mechanical and physical properties change and the liquid cannot be considered homogeneous. Usually when the solid particle velocity is less than 1 mm/s, the fluid is considered homogeneous. Several studies have been done but they give results for specific conditions i.e. absence of a universal method that expresses the influence of solid particles on the performance of centrifugal pumps.

The presence of solid particles in the flow tends to produce harmful effects on the performance of centrifugal pumps, for a reliable transport system and an optimal return detailed information about these effects is necessary. Although the presence of solid particles presents complicated effects than those that accompany an increase in viscosity, there is some qualitative similarity between the flow of rough mixes and liquid flow values of density and viscosity close as water. However, most research has been based on tests of centrifugal pumps handling industrial mixtures show that there are commonalities and equations found are semi empirical. [9]

Fairbank (1942) [10] presents a complete theoretical method which expresses the change in the characteristics of the pump according to the parameters of the mixture. The speeds of the solid particles and water to the output of the action wheel are calculated and used to calculate the height developed as a mixture.

$$H_m = \frac{w}{g \cdot \rho_m} [\, \rho_s. C_v(R_2.Cu_{,s} - R_1Cu_{1,o}) + (1-C_v)(R_2Cu_{2,o} - R_1Cu_{1,o})]$$

Where: ρ_m ρ_s are respectively mixture and solid densities.
and

$$\rho_m = \rho_s.C_v + (1 - C_v)$$

The diameter of the solid particles is 0.8 mm and the volume concentration of 20%.

Frasier [11] presents a simple expression for determining the coefficient of reduction of pump performance but does not include the dimensions of the solid particles and gives the expression for calculating the coefficients of height reduction and performance by:

$$K_\eta = K_H = \frac{1 - C_v}{1 - C_v + \varpi_s.C_v}$$

ϖ relative density of the solid

Vocaldo and Sagoo (1974) [12] simply suggest that the delivery head in a mixture of the pump decreases from 10 to 20% compared to that obtained in water.

McElvain (1976) [13] proposed that for a centrifugal pump with a wheel diameter of 350 mm the coefficient of height reduction and performance are equal and are directly proportional to the volume concentration of the mixture:

$$K_H = K_\eta = K_v \cdot \left(\frac{C_v}{0,20} \right)$$

- Coefficient between 0.05 for fine particles (10 microns) and 0.42 for particles having a diameter up to 10 mm. The curves of the coefficient K were compared with data from other researchers by Addie and Sellgren (1989) [14] which included their own data for large pumps examined at GIW Hydraulic Laboratory.

Reizes and Burgess (1976) [15] provide a method for estimating the characteristic of the pump in combination with knowledge of water and offer a clear empirical relationship between height and water mixture:

$$K_H = (1 - C_P)^n$$

K_H - Coefficient of height reduction

C_P - Concentration by weight

The value of parameter for each solid used is determined from Table 1

Table.1. *Value of n for each solid used*

	Type of solide	n
01	Beach sand	0.333
02	River sand	0.589
03	Ilménite	0.450
04	Heavy crude ore	0.561

Smolderev (1980) [16] shows that the height of the pump with concentrations lower than critical can be expressed as follows:

$$H_m = H_o \left(1 + K' \left(\frac{\rho_m}{\rho_o} - 1 \right)^n \right)$$

with $\qquad\qquad K' = 0,5 - 0,6$ et n $= 0,85$

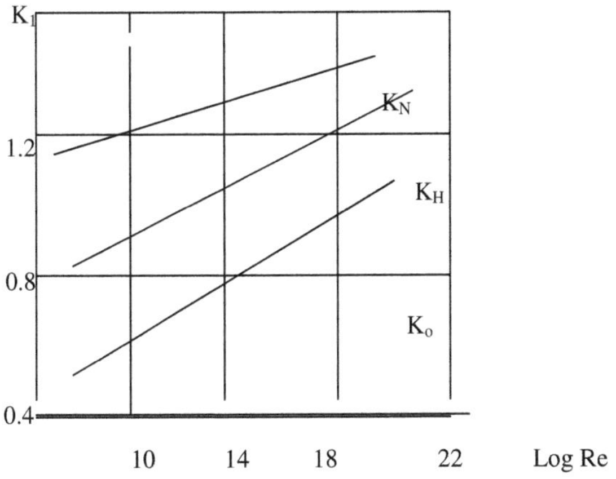

Figure.12. *Variation of the recalculation coefficients of the pump parameters depending on the flow regime [34]*

Moguilevski (1972) [17] proposes that the height of mixing created by the pump in the area of ± 20% of rated speed is:

$$H_m = H_o \frac{\rho_m}{\rho_o}$$

The Institute of Hydraulic Research of Ukraine [18] gives the relationship of height based on the mixture density and diameter of solid particles.

$$H_m = H_o \left(1 + \frac{0,5}{\psi^{0,5} + 0,5} \right) \left(\frac{\rho_m}{\rho_o} - 1 \right)$$

ψ - coefficient which takes into account the form of solid particles.
The equations of power and efficiency are expressed by:

$$N_m = N_O \frac{\rho_m}{\rho_o}$$

$$\eta_m = \eta_o \frac{\rho_m}{\rho_o}\left[1 + \frac{0,5}{\psi^{0.5}+0,5}\left(\frac{\rho_m}{\rho_o}-1\right)\right]$$

These relationships do not account for geometric and kinematic parameters of the pumps; they are only valid for pumps tested and with solid particles diameter between 0.05 and 0.1 mm.

Gahlot. V.K, V. Seshadri and R.C.Malhorta (1992) [19] using the influence of density, size and concentration on the performance of pump offers a reduction coefficient of the characteristic of the pump:

$$K = 0,00056(\bar{\omega}-1)^{0.72}\left(1+\frac{3}{\bar{\omega}}\right)C_p.Log\left(50.d\right)$$

d- Diameter of solid particles.

$\bar{\omega}$-- Relative density of the solid

This equation is similar to that proposed by Cave1976 [20], the difference results in the value of the exponent ($\bar{\omega}$ - 1) which takes into account the density of the solid

Practical solution as we can use the method proposed by Geoff moore (2003) which gives the relationship between the particle size of the solid particles, the volume concentration of the mixture, the ratio between the diameter of the solid particles and the diameter of the pump impeller and the coefficients of the head and effeciency. figure.13

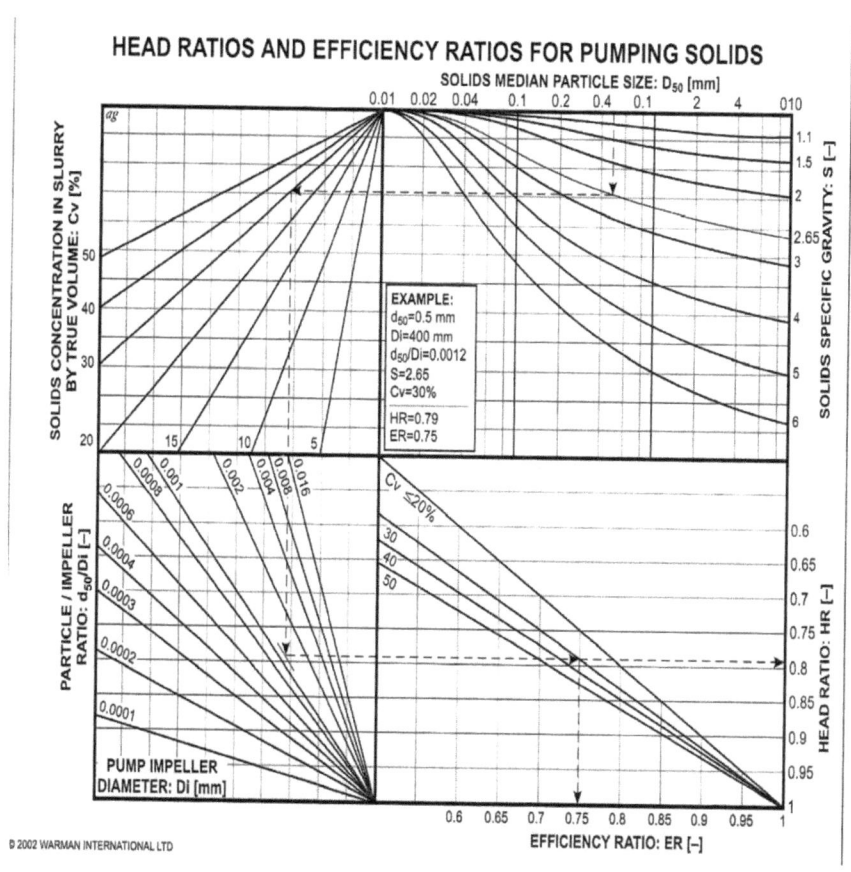

Figure.13 coefficients of head and performance according to the characteristics of the mixture and the solid [26]

4. Case study

The pump used in the test rig is a single stage centrifugal pump driven by an electric motor with a rated speed of 2900 rpm. To keep the same transport conditions during the tests, a closed circuit was used. The pump position facilitates the regulation of the flow and the mixture concentration without cavitations mode. The pump characteristics and solid particles are given in Figure.14.a, b and Table 2, 3. [25]

Table 2: Pump characteristics

Designations	value
Suction pipe (mm)	0.062
Control of repression (mm)	0.062
Diameter of the wheel (mm)	162
Type of wheel	Closed
A number of blade	6
Specific speed	70
Number of revolutions rpm	2900
Head of discharge (m)	30
Flow (1/s)	5.5

Table 3: Solid particle properties

Proprieties	Larry of washed phosphate
Density kg/ m^3	2800
Maximum diameter of the particles (mm)	0.2
Concentration %	30

.

By using the factors with undefined sizes of the head δ which is defined by: $\delta = g.H/(\omega^2.D^2)$ and of the flow σ defined by: $\sigma = Q/(U.D^3)$ or g is the acceleration of gravity in m/s², H is the head developed by the pump is expressed in meters of water column or in meters of column of mixture, Q the flow in m³/s, U the driven speed in m/s, ω the angular velocity in rad/s and D the diameter of the pump impeller in meters. Figure.15 a,b,c shows clearly that the concentration or the mixture density has a significant effect on the centrifugal pump performances. The decrease of head and efficiency increases with concentration in volume increasing. This reduction is more significant for higher concentrations.

The obtained curves reveal that the absolute decrease of the head H_o- H_m or efficiency

η_o - η_m for the same concentration is constant independently of the flow.[25]

Figure.14.a The centrifugal pump K-20-30

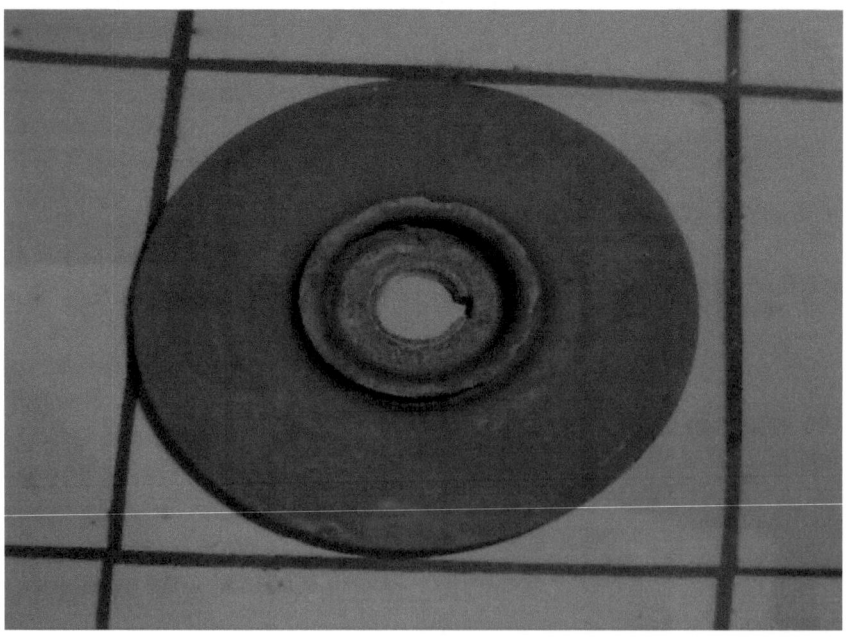

Figure.14.b The impeller of the centrifugal pump K-20-30

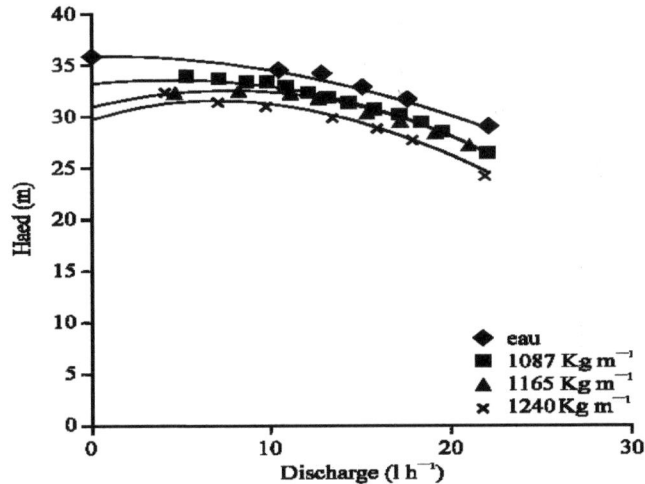

Figure.15a *Characteristic of head versus discharge for different mixture densities*

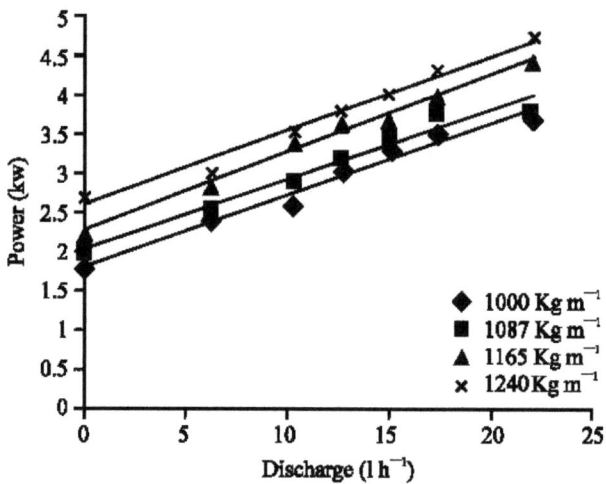

Figure.15.b *Characteristic of power versus discharge for different mixture densities*

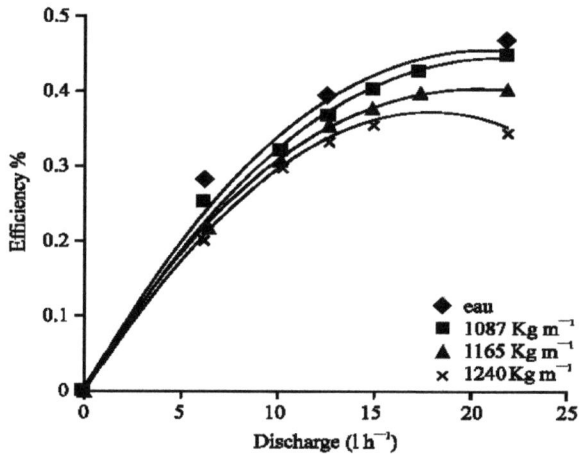

Figure.15.c *Characteristic of efficiency versus discharge for different mixture densities*

5. Material selection of the pump's components

Currently, a definition of abrasiveness does not exist but it is related to the hardness of the solid, to the nature of the contact between the solid particles and the pump parts and to the characteristics of the materials from which the pump is constructed. Table 4 shows some of the most common pump construction materials as a function of the hardness of the transported rock particles. It is observed that the impeller shape influences the characteristics of the pump working with solid–liquid mixtures. For an open pump impeller handling a mixture of a given concentration, efficiency (η) decreases by 4–6% and power (P) increases by 5–8% from the values for the same impeller pumping water (Ref. 10). In comparison, for a closed impeller pump, efficiency decreases by 20–30% and power increases by 20–30% for the same mixture concentration. According to the results obtained (Table 5), it can be seen that open impellers are better adapted to the transportation of solid–liquid mixtures [23]

Table.4 Appropriate pump construction materials for different hardness

Hardness	Transported material	Pump material
Hard	Topaz, quartz, feldspar	Iron resistant to wear
	Apatite	Treated steel, stainless steel
	Calcite, asbestos	Stainless steel, manganese
	Bauxite	Iron grey, stainless steel
Soft	Anthracite, alumina, gypsum plaster, sulphur	Plastic

Table.5. Comparison of the centrifugal pump results for different impellers [24]

Pump type	Impeller diameter (mm)	No of passages	Impeller type	Speed (rpm)	% with respect to water	
					η	P (kW)
1	24	5	Open	860 - 1600	-4 to -6	+5 to +8
2	24	2	Closed	950 - 1550	-20to -30	+20 to 35
3	30	2	Closed	850 - 1160	-5	+7.5
4	25	2	Closed	1000 - 1600	-4 to -6	+1 to +3

6. Conclusion

The theoretical study carried out in this chapter has allowed to demonstrate that the performance of pumps operating in mixture change compared to that of water, as well as the appearance of additional pressure drop due to the presence of solid particles. It should be noted that:

• changing the rotational speed leads to change the value of the critical diameter which in turn leads to changes in the percentage of large solid particles i.e. increase additional losses.

• Increased of additional losses in the wheel causes the decrease of the pressure head created by the pump.

• The speed of the solid particles inside the pump is a very important parameter for the calculation of the characteristics of centrifugal pumps and to evaluate the erosion of moving parts.

• Methods of the height correction, efficiency and power presented in the literature are determined for well-defined working conditions.

References

[1] Miguel Asuaje." Méthodologie et optimisation dans la conception des performances des turbomachines à fluide incompressible". Thèse d'état .ENSAM .centre de Paris 2003

[2] J.L.Kueny. EPFL/ LMH Prof. INPG." Turbomachines hydrauliques". Section de Genie Mécanique 2003

[3] Mikhailov.A "Les pompes à aubes." Moscou 1977.

[4] Joumakhov. J." Pompes, ventilateurs et compresseurs." Moscou 1970.

[5] T.S .Luu, B. Viney and L.Bencherif. "Turbomachine blading with splitter blades designed by solving the inverse flow field problem" J.Phys. III.france2. 657 -672. 1992.

[6] Gueijer.V Boroumenski."Hydraulique et commande hydraulique." Ed NEDRA 1982.

[7] Comolet .R" Mécanique expérimentale des fluides." Tome 2 ed Masson 1976

[8] Givotovski.L Smoylovskaia L "Les pompes à aubes pour mélanges abrasifs." Moscou. 1978

[9]K.C.Wilson, G.R.Addie, A .Sellgren : "Slurry transport using centrifugal pumps." Blackie Academic Professional.. Second édition. London 1997]

[10] William .A.Hunt , Robert.R.Faddick "The effects of solids on centrifugal pump characteristics . Advances in solid liquid flow in pipes and its applications. "London 1971

[11] I. Zandi." Advances in solid liquid flow in pipes and its qpplications. "Per;magnon paper by Hunt . Faddick. 1971.

[12] J.J Vocaldo and M.S Sagoo." Slurry flow in pipes and pumps. "Trans A.S.M.E vol 95 series B Journal of Engineering for industry N°1 Feb 1973 p65.

[13]McElvain .R.E." High pressure pumping .Skillings."Mining Review, 63 (4) pp 1-14 1974.

[14] Sellgren. A and Addie. G.R." Effect of solids on large centrifugal pump head and efficiency." Paper presented at CEDA Dredging Day. Amsterdam The Netherlands 1989.

[15]K.E.Burgess andA. Reizes " The effect of sizing specific gravity and concentration on the perfermance of centrifugal pumps." Proc Inst Mech Eng vol 190 (36) 391-399 1976.

[16]. Smolderev " Le transport par conduites." Moscou. 1980.

[17]I. Moguilevski . "Essais des pompes à sable pour les suspensions lourdes." .Moscou cnitihimmach .1972.

[18]B. Karacik,U. Assaulenko. "Transport hydraulique du sable." Kiev; Naoukova-Doumka 1966.

[19]Gahlot . «Effects of densite ,size distribution and concentration of solid on the characteristics of centrifugal pumps." Trans A.S.ME vol 114 Journal of Fluids Engineering 1992 p386.

[20]. Cave." Effects of suspended solids on the performance of centrifugal pumps." Proc Hydro transport4 paper43. 1976.

[21] Benretem,A., Haddouche,A., Cheghib,H., and Saad,S., "Influence of Solid Particles on Centrifugal PumpCharacteristics,"Journal of Engineering and Applied Sciences, Medwell Journals, 2007.

[22] A Benretem,D Khalfa and M Benidir   Detremination of centrifugal pump characteristics in solid liquid mixture World Journal of Engineering 4(4)2071-6

[23] A.Benretem, M.C.Benidir and R.Chaib Mixture type dictates pump choice Volume 2009, Issue 510. pp 32-34 Wold Pumps Volume 2009, Issue 510

[24] W.A. Hunt and R.R. Faddik, The effect of solids on centrifugal pump characteristic, Advances in solid liquid flow in pipes and its applications, (I.Zandi, ed),pp.271-278,Pergamon Press,(1971).

[25] A.Benretem; A.Hadouche , H. Cheghib and S.Saad . Influence of solid particles on centrifugal pump characteristics.Journal of Engineering and Applied Science 2(1) 244-247, 2007

[26] Geoff.Moore, Hydraulic Conveying of Bulk Materials, Excellent Engineering Solutions. Weir Minerals Division. USA Mars 2003.

Printed by Books on Demand GmbH, Norderstedt / Germany